Understanding Monkeypox: A Comprehensive Guide to the Emerging Viral Threat

"What It Is, How It Spreads, and How to Protect Yourself and Your Community"

Dr. Ava Sinclear

© 2024, DR. AVA SINCLEAR

All rights reserved. No part of this publication may be reproduced, distributed, or transmitted in any form or by any means, including photocopying, recording, or other electronic or mechanical methods, without the prior written permission of the author, except in the case of brief quotations embodied in critical reviews and certain other noncommercial uses permitted by copyright law.

DEDICATION

To those who work tirelessly on the frontlines of public health centres and other private clinics or healthcentres, combating infectious diseases with courage and determination.

And to every individual who strives to stay informed and protect their families and communities your vigilance and care are the first line of defence.

This book is for you!

TABLE OF CONTENTS

1. Introduction to Monkeypox

2. What is Monkeypox?

3. The Transmission of Monkeypox

4. Symptoms and Diagnosis

5. Treatment and Management

6. Prevention Strategies

7. Global and Local Responses

8. The Future of Monkeypox

9. Conclusion

THANK YOU!

INTRODUCTION TO MONKEYPOX

The History and Origins of Monkeypox

Monkeypox was first discovered in 1958 when outbreaks occurred in monkeys kept for research. However, the first human case wasn't recorded until 1970 in the Democratic Republic of the Congo. Since then, most cases have been reported in Central and West Africa, especially in remote areas where people may come into close contact with wild animals. The virus is thought to be carried by rodents, such as rats and squirrels, which can pass it on to other animals and humans.

WHY UNDERSTANDING MONKEYPOX MATTERS TODAY

Monkeypox is important to understand because it can spread from animals to humans and from person to person, leading to outbreaks. Although it is not as contagious or deadly as smallpox, it still poses a serious health risk, especially in areas with limited access to healthcare.

With recent cases reported outside of Africa, it's crucial to know how the disease spreads and how to protect ourselves and our communities. Understanding monkeypox helps us stay informed, take preventive measures, and also respond effectively if an outbreak occurs.

WHAT IS MONKEYPOX?

Overview of the Monkeypox Virus

Monkeypox is a viral infection caused by the monkeypox virus, which belongs to the Orthopoxvirus genus, the same group of viruses that includes smallpox. While monkeypox is less severe than smallpox, it is still a significant disease that can lead to serious health complications. The virus can infect both humans and animals and is primarily found in Central and West Africa, particularly in areas close to tropical rainforests where animals that may carry the virus are common.

Monkeypox is a zoonotic disease, meaning it can be transmitted from animals to humans. The virus is believed to be naturally harboured in certain wild animals, especially rodents like rats and squirrels.

Occasionally, these animals can transmit the virus to humans, either directly through bites or scratches or indirectly through contact with their bodily fluids or contaminated materials. Once the virus infects a human, it can also spread from person to person through respiratory droplets, close contact with an infected person's skin lesions, or by touching contaminated objects like bedding or clothing.

Symptoms and Stages of the Disease

Monkeypox symptoms typically appear within 5 to 21 days after exposure to the virus. The disease progresses through several stages, and its symptoms can be divided into two main phases:

1.INITIAL PHASE (Prodromal Stage)

- **Fever:** One of the first symptoms is a high fever, which may be accompanied by chills.
- **Headache:** Intense headaches are common during the early stage of infection.
- **Muscle Aches:** Patients often experience muscle pain, along with backaches.
- **Swollen Lymph Nodes:** Unlike smallpox, monkeypox causes noticeable swelling of the lymph nodes (lymphadenopathy), particularly in the neck,

armpits, and groyne. This is a key distinguishing feature of the disease.

•**Fatigue:** Generalised weakness and exhaustion are common during this phase.

2. RASH PHASE (Eruptive Stage)

•**Development of Rash:** A few days after the fever begins, a rash starts to develop, usually on the face first, and then it spreads to other parts of the body, including the hands, feet, chest, and sometimes the genital area and eyes.

THE RASH PROGRESSES THROUGH SEVERAL STAGES

•**Macules:** Flat, discoloured spots on the skin.
•**Papules:** Raised bumps that develop on the macules.
•**Vesicles:** Small fluid-filled blisters form on the papules.
•**Pustules:** The vesicles become pus-filled, turning into pustules.
•**Scabs:** Eventually, the pustules dry out and form scabs, which later fall off, leaving scars in some cases.
•**Duration of Rash:** The rash phase typically lasts for about 2 to 4 weeks. The number of lesions can

vary from a few to several thousand, and the severity of the rash can differ from person to person.

ADDITIONAL SYMPTOMS

 •**Respiratory Symptoms:** Some patients may develop respiratory symptoms, such as a cough or sore throat, particularly if the virus spreads through respiratory droplets.
 •**Gastrointestinal Symptoms:** In some cases, nausea, vomiting, and diarrhoea may occur, especially if the virus affects the digestive system.

•**Severity and Complications:**
While monkeypox is generally self-limiting, meaning it resolves on its own without specific treatment, it can be severe in some cases, especially in children, pregnant women, and individuals with weakened immune systems. Complications can include secondary bacterial infections, pneumonia, and in rare cases, encephalitis (inflammation of the brain), which can be life-threatening.

Understanding the symptoms and stages of monkeypox is crucial for early diagnosis and effective management of the disease. Prompt recognition of the signs can help prevent the spread of the virus and reduce the risk of complications.

THE TRANSMISSION OF MONKEYPOX

Monkeypox is a zoonotic disease, which means it primarily spreads from animals to humans. However, it can also spread from person to person. Understanding the transmission pathways is crucial for preventing outbreaks and protecting public health.

How Monkeypox Spreads from Animals to Humans

Monkeypox is thought to be naturally maintained in certain wild animals, particularly rodents like squirrels, rats, and dormice. These animals are considered the primary reservoirs of the monkeypox virus. While monkeys can also carry the virus, they are not the main reservoir but rather accidental hosts, which is how the virus got its name.

1. DIRECT CONTACT WITH INFECTED ANIMALS

- The most common way monkeypox spreads to humans is through direct contact with an infected animal. This can occur when a person is bitten or scratched by an infected animal or when they handle the animal's blood, bodily fluids, or lesions. For example, people who hunt wild animals for food, known as "bushmeat," or those who keep exotic pets, may be at higher risk of contracting the virus.
- Handling animals that have died from monkeypox or consuming undercooked meat from infected animals can also lead to transmission. The virus can enter the body through broken skin, even if the skin is not visibly damaged, or through mucous membranes, such as the eyes, nose, or mouth.

2. INDIRECT CONTACT WITH CONTAMINATED MATERIALS

- The virus can also spread indirectly through contact with objects that have been contaminated by an infected animal. For example, bedding, clothing, or other materials that have come into contact with an infected animal's lesions or bodily fluids can carry the virus. When a person touches these contaminated items, they can become infected.

- In areas where monkeypox is endemic, people may come into contact with contaminated materials while farming, hunting, or engaging in other activities that bring them close to infected animals.

Human-to-Human Transmission: Risks and Precautions

While animal-to-human transmission is the primary way monkeypox spreads, human-to-human transmission is also possible, though it is less common. Understanding how the virus spreads between people is vital for controlling outbreaks, particularly in settings where people live in close quarters or have frequent physical contact.

1. RESPIRATORY DROPLETS

- The most significant mode of human-to-human transmission is through respiratory droplets. When an infected person coughs, sneezes, or talks, they release droplets into the air that can carry the virus. These droplets can then be inhaled by someone nearby, leading to infection.
- Transmission through respiratory droplets generally requires prolonged face-to-face contact, making household members, healthcare workers, or others in close contact with an infected person

particularly vulnerable. This is why monkeypox outbreaks can spread more rapidly in enclosed environments, such as homes, hospitals, or other settings where people are in close proximity.

2. DIRECT CONTACT WITH SKIN LESIONS

• Another way monkeypox spreads between people is through direct contact with the skin lesions of an infected person. During the rash phase of the disease, the blisters and sores contain a high concentration of the virus. Touching these lesions, or coming into contact with the fluids within them, can result in transmission.

• This type of transmission can occur during physical contact, such as hugging, kissing, or sexual contact. Caregivers and healthcare workers are at increased risk if they do not use proper protective measures when treating or caring for infected individuals.

3. CONTACT WITH CONTAMINATED OBJECTS

• Similar to the indirect transmission from animals, monkeypox can spread between people through contact with objects that have been contaminated by an infected person's lesions or bodily fluids. This includes items like bedding, towels, clothing, or personal items.

• In healthcare settings, the virus can spread through medical equipment or surfaces that have not been properly disinfected. Ensuring that contaminated materials are handled and cleaned appropriately is critical to preventing further transmission.

4. VERTICAL TRANSMISSION

• Vertical transmission, or mother-to-child transmission, can occur during pregnancy if the virus crosses the placenta. This can lead to congenital monkeypox, which is associated with severe outcomes for the newborn.

Precautions to Prevent Human-to-Human Transmission

•**Isolation:** Infected individuals should be isolated from others to prevent the spread of the virus. This includes staying in a separate room and avoiding contact with others as much as possible until the rash has healed and all scabs have fallen off.

•**Use of Personal Protective Equipment (PPE):** Healthcare workers and caregivers should wear appropriate PPE, including gloves, masks, and gowns, when caring for infected individuals. This helps to minimise the risk of transmission.

•Hygiene Practices: Regular handwashing with soap and water is essential, especially after contact with an infected person or contaminated materials. Disinfecting surfaces and laundering contaminated bedding and clothing in hot water can also help reduce the risk of spread.

•**Vaccination:** The smallpox vaccine has been shown to provide some protection against monkeypox. In some cases, vaccination may be recommended for those at high risk of exposure, such as healthcare workers or close contacts of an infected person.

Case Studies of Recent Outbreaks

Monkeypox was largely confined to Central and West Africa until the early 2000s when cases began to emerge in other regions, highlighting the potential for global spread.

1. THE 2003 U.S. OUTBREAK

• **The first monkeypox** outbreak in the United States occurred in 2003 and was linked to the importation of infected rodents from West Africa.

These animals were housed with prairie dogs, which were then sold as pets. The virus spread to humans who had close contact with these pets, leading to over 70 confirmed and suspected cases. Fortunately, no deaths were reported, but the outbreak underscored the risks of the global exotic pet trade in spreading zoonotic diseases.

2. THE 2017-2018 NIGERIAN OUTBREAK:

- Nigeria experienced a significant monkeypox outbreak between 2017 and 2018, with over 300 suspected cases and several deaths. This outbreak was notable because it affected urban areas, where human-to-human transmission played a significant role in the spread of the virus. The Nigerian outbreak highlighted the challenges of controlling monkeypox in densely populated areas and the importance of rapid response and public health measures.

3. THE 2021-2022 GLOBAL OUTBREAKS

- **In 2021 and 2022,** monkeypox cases were reported in several countries outside of Africa, including the United Kingdom, the United States, and several European nations. These outbreaks were linked to international travel and close contact with infected individuals. The global spread of monkeypox during this period raised concerns about the virus's potential to cause widespread outbreaks, leading to increased surveillance, public health campaigns, and discussions about the need for vaccination strategies.

These case studies demonstrate that while monkeypox has historically been a localised disease, it has the potential to spread globally, particularly with increased travel and trade. This makes it essential to monitor the virus and take preventive measures to reduce the risk of future outbreaks.

Monkeypox transmission occurs primarily through contact with infected animals but can also spread from person to person through respiratory droplets, direct contact with skin lesions, and contaminated objects.

Understanding these transmission pathways and implementing precautions, such as isolation, hygiene practices, and vaccination, is crucial for controlling outbreaks and protecting public health. The recent outbreaks outside of Africa serve as a reminder that monkeypox is not just a regional concern but a global health issue that requires continued vigilance and preparedness.

SYMPTOMS AND DIAGNOSIS

Monkeypox is a viral infection with symptoms that can vary in severity, ranging from mild to potentially life-threatening, especially in vulnerable populations. Understanding the symptoms and how the disease progresses is crucial for early detection and effective treatment. Accurate diagnosis is also essential for managing outbreaks and preventing the spread of the virus.

Identifying Early Symptoms

The early symptoms of monkeypox are often similar to those of other viral infections, which can make initial diagnosis challenging without proper testing. However, certain key symptoms can help identify the disease.

1. INCUBATION PERIOD
  • The incubation period, or the time from exposure to the virus to the onset of symptoms, typically ranges from 5 to 21 days, with most cases occurring within 7 to 14 days. During this period,

the virus is replicating in the body, but no symptoms are yet present.

2. PRODROMAL (Early) SYMPTOMS:
   •**Fever:** The first noticeable symptom of monkeypox is usually a sudden onset of fever. The fever can be high and is often accompanied by chills.
   •**Headache:** Intense headaches are common during the early stage of the disease.
   •Muscle Aches and Back Pain: Patients frequently experience muscle aches (myalgia) and back pain, which can be severe.
   •**Fatigue and Exhaustion:** Generalised weakness and a feeling of extreme tiredness are typical during the prodromal stage.
   •**Swollen Lymph Nodes (Lymphadenopathy):** One of the distinguishing features of monkeypox, compared to similar diseases like smallpox, is the presence of swollen lymph nodes. Lymphadenopathy typically occurs in the neck, armpits, or groyne and can be painful. This symptom is significant because it can help differentiate monkeypox from other pox-like illnesses.

These early symptoms usually last for 1 to 3 days before the appearance of the characteristic rash, which marks the next phase of the disease.

The Progression of the Disease

As monkeypox progresses, the symptoms become more specific, with the development of a rash being the most prominent feature. The disease can be divided into several stages based on the appearance and evolution of the rash.

1. RASH DEVELOPMENT:

The rash usually begins within a few days after the onset of fever and other prodromal symptoms. It typically starts on the face and then spreads to other parts of the body, including the hands, feet, chest, and sometimes the genital area and eyes. The rash may also appear on the palms of the hands and the soles of the feet, which is less common in many other viral infections.

- **The rash progresses through several stages**

- **Macules:** The initial lesions are flat, discoloured spots on the skin, known as macules. These macules are typically small and may be red or darker than the surrounding skin.

- **Papules:** Within a day or two, the macules raise and become papules, which are small, firm bumps.
- **Vesicles:** The papules then fill with clear fluid, forming vesicles. These vesicles are small, blister-like lesions.
- **Pustules:** The vesicles become pustules as they fill with pus. Pustules are typically round, firm, and often umbilicated (having a central depression). This stage is often the most uncomfortable, as the pustules can be painful and itchy.
- **Scabs:** After about 5 to 7 days, the pustules begin to dry out and form scabs. The scabs eventually fall off, and the skin underneath may be lighter or darker for a time. In some cases, the lesions can leave permanent scars, especially if they become infected.

2. ADDITIONAL SYMPTOMS

- In addition to the rash, some patients may experience respiratory symptoms, such as a cough, sore throat, or shortness of breath, particularly if the virus spreads through respiratory droplets.
- Gastrointestinal symptoms like nausea, vomiting, and diarrhoea may also occur, especially if the virus affects the digestive system.

3. DURATION OF THE DISEASE

• The entire course of the disease typically lasts 2 to 4 weeks. In most cases, the disease is self limiting, meaning it resolves on its own without the need for specific antiviral treatment. However, in severe cases, particularly in immunocompromised individuals, children, or pregnant women, complications can arise, such as secondary bacterial infections, pneumonia, sepsis, or encephalitis (inflammation of the brain).

4. SEVERITY AND COMPLICATIONS

• The severity of monkeypox can vary widely. While many cases are mild, some can be severe, particularly in individuals with weakened immune systems. Complications, although rare, can include dehydration, secondary bacterial infections (which can occur if the skin lesions are not properly cared for), and respiratory distress. In some instances, monkeypox can be fatal, especially in young children or individuals with compromised health.

Diagnostic Methods and Tools

Accurate diagnosis of monkeypox is essential for confirming cases, initiating appropriate treatment, and preventing further transmission. Diagnosis typically involves a combination of clinical evaluation and laboratory testing.

1. CLINICAL EVALUATION

• **Symptom Assessment:** The first step in diagnosing monkeypox is a thorough clinical evaluation, including a detailed assessment of symptoms. The presence of fever, swollen lymph nodes, and a characteristic rash are key indicators that may suggest monkeypox. Healthcare providers will also take into account the patient's travel history, possible exposure to infected animals or individuals, and any recent activities that might have put them at risk.

Differential Diagnosis: Because monkeypox shares symptoms with other diseases like chickenpox, smallpox, and other poxvirus infections, healthcare providers must carefully distinguish between these conditions. The presence of lymphadenopathy (swollen lymph nodes) is a helpful distinguishing feature, as it is less common in smallpox and chickenpox.

2. LABORATORY TESTING

- **Polymerase Chain Reaction (PCR):** The most reliable method for diagnosing monkeypox is PCR testing. This involves taking samples from the patient's lesions, such as fluid from vesicles or pustules, or swabs of the lesions, and testing them for the presence of monkeypox viral DNA. PCR is highly specific and can differentiate monkeypox from other poxviruses.
- **Virus Isolation:** In some cases, the virus can be isolated from a sample and grown in a laboratory setting. This method, while accurate, is time-consuming and requires specialised facilities, so it is less commonly used for routine diagnosis.
- **Serology:** Serological tests can detect antibodies against the monkeypox virus in the blood, indicating a current or past infection. However, serology is not always definitive, as cross-reactivity with other orthopoxviruses, such as the vaccinia virus (used in the smallpox vaccine), can occur.
- **Electron Microscopy:** In some specialised labs, electron microscopy can be used to visualise the virus directly from lesion samples. While not commonly used for routine diagnosis, it can provide confirmation in certain cases.

3. EPIDEMIOLOGICAL CONSIDERATIONS

- **Travel and Exposure History:** When diagnosing monkeypox, it is crucial to consider the patient's recent travel history, especially if they have visited regions where monkeypox is endemic, such as Central and West Africa. Additionally, any known exposure to infected animals, individuals, or materials should be taken into account.
- **Outbreak Investigation:** During an outbreak, public health authorities often perform additional epidemiological investigations to trace the source of the virus and identify potential contacts who may be at risk. This can involve testing of animals, environmental sampling, and contact tracing to prevent further spread.

Recognizing the symptoms of monkeypox and understanding the progression of the disease are critical for early detection and treatment. While the disease often begins with general symptoms like fever and fatigue, the distinctive rash and swollen lymph nodes can help distinguish monkeypox from other infections.

Laboratory testing, particularly PCR, plays a vital role in confirming the diagnosis and guiding public health responses. Early and accurate diagnosis not only helps in treating affected individuals but also in preventing the spread of monkeypox within communities.

TREATMENT AND MANAGEMENT

Monkeypox, while often self limiting, requires careful treatment and management to ensure a full recovery and to prevent complications, especially in severe cases.

This section will explore the current treatment options, the importance of supportive care, and strategies for managing potential complications.

Current Treatment Options for Monkeypox

There is no specific antiviral treatment that is universally approved for monkeypox, but several options can be used to manage the disease, particularly in severe cases or those at high risk for complications.

1. ANTIVIRAL MEDICATIONS

Tecovirimat (TPOXX): Tecovirimat is an antiviral medication that has been approved for the treatment of smallpox and has shown promise in treating monkeypox as well. It works by inhibiting the viral protein responsible for the release of the virus from infected cells, thereby preventing its spread within the body. Tecovirimat is typically considered for use in severe cases of monkeypox, especially in immunocompromised patients, pregnant women, children, and those with extensive lesions.

Cidofovir and Brincidofovir: These antiviral agents, originally developed for other viral infections, have also been used to treat monkeypox, though they are generally reserved for more severe cases. Cidofovir has been used in cases of progressive or severe monkeypox, particularly when tecovirimat is not available. Brincidofovir, a newer and less toxic version of cidofovir, has shown potential in treating orthopoxvirus infections, including monkeypox.

• **Vaccinia Immune Globulin (VIG):** VIG is an immune globulin product derived from the blood of individuals who have been vaccinated against smallpox. It can be used to treat complications from smallpox vaccination and has been considered for use in severe monkeypox cases, particularly in individuals who cannot take other antivirals.

2. SYMPTOMATIC TREATMENT

• While specific antiviral treatments are used in severe cases, most patients with monkeypox are treated with symptomatic care to relieve symptoms and prevent secondary infections.

Commonly used medications include

• **Antipyretics and Analgesics:** Medications such as acetaminophen or ibuprofen are used to reduce fever and relieve pain associated with the disease.

•**Antihistamines:** These can help alleviate itching associated with the rash, making patients more comfortable as the lesions heal.

•**Topical Antibiotics:** In some cases, topical antibiotics may be applied to skin lesions to prevent bacterial superinfection, especially if the lesions become secondarily infected due to scratching or other trauma.

3. VACCINATION

• **Smallpox Vaccine:** The smallpox vaccine, particularly the newer ACAM2000 and Jynneos vaccines, has been shown to provide some protection against monkeypox due to the similarity between the two viruses. In some cases, post-exposure vaccination (within 4 days of exposure) can help prevent the onset of the disease or reduce its severity. Vaccination is particularly recommended for healthcare workers, close contacts of infected individuals, and those at higher risk of severe disease.

Supportive Care and Recovery

Supportive care is a cornerstone of monkeypox management, as it helps alleviate symptoms, supports the immune system, and promotes recovery.

1. HYDRATION AND NUTRITION

• **Maintaining Hydration:** Patients with monkeypox, particularly those with high fever or gastrointestinal symptoms, may be at risk for dehydration. Ensuring adequate fluid intake is crucial, and in some cases, intravenous fluids may be required if the patient is unable to maintain oral intake.

• **Nutritional Support:** Proper nutrition is important for supporting the immune system and promoting healing. Patients should be encouraged to eat a balanced diet, and in cases of severe illness, nutritional supplements may be provided.

2. SKIN CARE AND WOUND MANAGEMENT

• **Caring for Skin Lesions:** Proper care of skin lesions is essential to prevent secondary infections and promote healing. Patients should be advised to keep the lesions clean and dry. In some cases, saline baths or the application of emollients can help soothe the skin and prevent cracking or infection.

• **Avoiding Scratching:** Patients should be advised not to scratch the lesions, as this can lead to scarring or secondary bacterial infections. Nails should be kept short, and if necessary, gloves or mittens can be worn, especially in children, to prevent scratching.

3. ISOLATION AND REST

• **Home Isolation:** Patients with monkeypox should be isolated from others to prevent the spread of the virus. They should remain at home until all lesions have crusted over, the scabs have fallen off, and a fresh layer of skin has formed. During this period, they should avoid close contact with others, particularly those who are immunocompromised.

- **Rest and Recovery:** Rest is essential for recovery. Patients should be encouraged to get plenty of sleep and avoid strenuous activities while they are symptomatic.

4. MONITORING FOR COMPLICATIONS

- **Regular Check-ups:** Regular medical check-ups are important, especially in severe cases, to monitor the patient's condition and detect any early signs of complications. Healthcare providers should monitor for signs of respiratory distress, secondary bacterial infections, or worsening skin lesions.

Managing Complications

While most cases of monkeypox are mild, complications can occur, particularly in vulnerable populations such as children, pregnant women, and individuals with weakened immune systems. Early recognition and management of these complications are vital for improving outcomes.

1. SECONDARY BACTERIAL INFECTIONS

- **Prevention and Treatment:** Skin lesions, especially if not properly cared for, can become secondarily infected with bacteria. This can lead to cellulitis, abscesses, or other skin infections. Topical or systemic antibiotics may be required to treat these infections. Patients should be advised on proper wound care and hygiene to prevent these complications.

2. RESPIRATORY COMPLICATIONS

- **Pneumonia:** In some cases, monkeypox can lead to respiratory complications, including pneumonia. This is more likely to occur in severe cases or in those with pre-existing respiratory conditions. Treatment may involve supportive care, such as oxygen therapy, and in some cases, antiviral or antibacterial treatment if secondary infection is suspected.
- **Bronchitis and Croup:** In children, monkeypox can lead to bronchitis or croup, which may require medical intervention, including the use of nebulizers or corticosteroids to reduce airway inflammation.

3. ENCEPHALITIS

- **Neurological Complications:** Encephalitis, or inflammation of the brain, is a rare but serious complication of monkeypox. Symptoms may include headache, confusion, seizures, and altered mental status. Treatment typically involves supportive care in a hospital setting, and in some cases, antiviral treatment may be considered.

4. OCULAR COMPLICATIONS

• **Eye Involvement:** Monkeypox lesions can sometimes affect the eyes, leading to conjunctivitis or more severe ocular complications, such as corneal ulcers or vision loss. Early treatment with antiviral eye drops or other interventions may be necessary to prevent long-term damage.

5. DEHYDRATION AND ELECTROLYTE IMBALANCE

• **Fluid Management:** Severe gastrointestinal symptoms, such as vomiting and diarrhoea, can lead to dehydration and electrolyte imbalances. Patients may require intravenous fluids and electrolyte replacement to maintain hydration and prevent complications such as kidney failure or shock.

The treatment and management of monkeypox involve a combination of antiviral therapy, symptomatic treatment, and supportive care. While most cases are self-limiting and resolve with proper care, some patients may experience severe disease or complications that require more intensive medical intervention. Early recognition, proper management of symptoms, and timely treatment of complications are key to ensuring a full recovery and preventing the spread of monkeypox within communities. As the understanding of monkeypox continues to evolve, ongoing research and public health efforts will be essential in improving treatment options and outcomes for those affected by this disease.

PREVENTION STRATEGISTS

Preventing monkeypox requires a multi-faceted approach that addresses the different ways the virus can spread. Effective prevention strategies involve reducing the risk of animal-to-human transmission, implementing personal and community-based measures, and utilising vaccination to protect at-risk populations.

Each of these strategies plays a crucial role in controlling and eventually eliminating monkeypox outbreaks.

Preventing Animal-to-Human Transmission

Monkeypox is a zoonotic disease, meaning it can be transmitted from animals to humans. Preventing this form of transmission is critical, especially in areas where the virus is endemic in animal populations.

1. AVOIDING CONTACT WITH INFECTED ANIMALS

- **Wildlife Interaction:** The primary way monkeypox is transmitted from animals to humans is through direct contact with infected animals, particularly rodents, primates, and other small mammals that serve as reservoirs for the virus. To prevent infection, people should avoid handling wild animals, especially those that are sick or found dead. Hunting, capturing, or consuming bushmeat (wild game) should be discouraged, particularly in regions where monkeypox is known to circulate.
- **Domestic Animals:** In some cases, domestic animals can become infected with monkeypox if they come into contact with wild animals or contaminated environments. It's important to monitor pets for signs of illness and prevent them from interacting with potentially infected wildlife.

2. SAFE HANDLING AND PREPARATION OF ANIMAL PRODUCTS

- **Cooking and Food Safety:** If wild animals are consumed, it is essential that the meat is thoroughly cooked to kill any potential pathogens. Raw or undercooked meat poses a significant risk of transmission, not just of monkeypox but of other zoonotic diseases as well.

• **Protective Measures for Animal Handlers:** Individuals who work with animals, such as farmers, veterinarians, and wildlife handlers, should wear protective clothing, including gloves and masks, when dealing with animals that could be infected. They should also follow strict hygiene practices, such as washing hands and disinfecting equipment after handling animals or animal products.

3. CONTROLLING ANIMAL POPULATIONS

• **Rodent Control:** Since rodents are a key reservoir for the monkeypox virus, controlling rodent populations around homes, farms, and other human habitats is vital. This can involve measures such as sealing entry points to buildings, removing food sources, and using traps or rodenticides in areas where rodents are prevalent.

• **Animal Quarantine:** In regions where monkeypox is endemic, it may be necessary to quarantine and monitor animals that are suspected of being infected. This can help prevent the spread of the virus to other animals or humans.

Personal and Community Prevention Measures

Beyond preventing transmission from animals, it's important to implement measures that reduce the risk of person-to-person transmission and protect communities during outbreaks.

1. PERSONAL PROTECTIVE MEASURES

- **Hand Hygiene:** Regular hand washing with soap and water is one of the most effective ways to prevent the spread of monkeypox. If soap and water are not available, an alcohol-based hand sanitizer can be used. This is particularly important after contact with potentially contaminated surfaces or materials, such as bedding, clothing, or medical equipment.
- **Use of Personal Protective Equipment (PPE):** During an outbreak, healthcare workers and those caring for infected individuals should wear appropriate PPE, including gloves, masks, and gowns, to prevent exposure to the virus. This is especially important when dealing with bodily fluids or contaminated materials.
- **Avoiding Close Contact:** People should avoid close physical contact with individuals who are suspected or confirmed to have monkeypox. This includes refraining from activities such as hugging, kissing, or sharing personal items like towels, bedding, or utensils with an infected person.

2. COMMUNITY PREVENTION EFFORTS

- **Public Awareness and Education:** Educating communities about the risks of monkeypox and how to prevent its spread is essential, especially in regions where the virus is endemic or where an outbreak has occurred. Public health campaigns can help inform people about the symptoms of monkeypox, how it spreads, and what steps they can take to protect themselves and others.
- **Isolation and Quarantine:** In the event of an outbreak, isolating infected individuals and quarantining those who have been exposed can help prevent further spread. Isolation should continue until the infected person is no longer contagious, typically after all lesions have scabbed over and fallen off. Public health authorities should provide support to ensure that those in isolation or quarantine can access food, medical care, and other necessities.
- **Environmental Disinfection:** Surfaces and objects that may have been contaminated by an infected person, such as bedding, clothing, and medical equipment, should be thoroughly disinfected. Cleaning with standard household disinfectants or bleach can kill the monkeypox virus and reduce the risk of transmission.

3. TRAVEL PRECAUTIONS

- **Travel Advisory:** Individuals travelling to regions where monkeypox is endemic or where an outbreak is occurring should take extra precautions, such as avoiding contact with animals and practising good hygiene. Public health authorities may issue travel advisories to inform travellers of specific risks and recommended preventive measures.
- **Monitoring and Reporting Symptoms:** Travellers returning from areas with monkeypox outbreaks should monitor themselves for symptoms and seek medical attention if they develop fever, rash, or other signs of the disease. Early detection and reporting of cases are crucial for controlling the spread of the virus.

The Role of Vaccination

Vaccination is a powerful tool in preventing monkeypox, particularly in high-risk populations and during outbreaks. The smallpox vaccine, due to its cross-protection against monkeypox, plays a key role in this strategy.

1. SMALLPOX VACCINE (ACAM2000 and Jynneos)

- **Cross-Protection:** The smallpox vaccine has been shown to provide some protection against monkeypox because both viruses belong to the orthopoxvirus family. Two vaccines, ACAM2000 and Jynneos, are currently available for use against monkeypox. Jynneos, also known as Imvamune or Imvanex, is a newer vaccine that is non-replicating, meaning it is safer for use in people with weakened immune systems or other contraindications to live vaccines.
- **Vaccination Strategy:** In regions where monkeypox is endemic or during an outbreak, vaccination can be used as a preventive measure. This may involve vaccinating healthcare workers, close contacts of infected individuals, and others at high risk of exposure. In some cases, post-exposure vaccination (within 4 days of exposure) can help prevent the onset of the disease or reduce its severity.

2. Pre-Exposure Vaccination

- **High-Risk Groups:** Certain groups of people, such as laboratory workers who handle orthopoxviruses, healthcare workers in outbreak areas, and individuals living in or travelling to endemic regions, may be recommended for pre-exposure vaccination. This can help protect them from infection if they come into contact with the virus.

3. POST-EXPOSURE PROPHYLAXIS (PEP)

• **PEP for Close Contacts:** For individuals who have been exposed to monkeypox, receiving a smallpox vaccine within a few days of exposure can reduce the likelihood of developing the disease or lessen its severity. This strategy is particularly useful during outbreaks to control the spread of the virus.

4. PUBLIC HEALTH CONSIDERATIONS

• **Vaccine Availability and Distribution:** Ensuring that vaccines are available and accessible in areas where they are needed most is crucial for preventing monkeypox. Public health authorities must coordinate efforts to distribute vaccines, especially in low-resource settings or during large-scale outbreaks.

Preventing monkeypox requires a comprehensive approach that includes reducing animal-to-human transmission, implementing personal and community-level preventive measures, and utilising vaccination to protect those at risk. By combining these strategies, it is possible to control and eventually eliminate monkeypox outbreaks, safeguarding public health and preventing the spread of this potentially serious disease. Ongoing research, public health initiatives, and education are key to improving prevention efforts and ensuring that communities are prepared to respond effectively to monkeypox.

GLOBAL AND LOCAL RESPONSES

The global and local responses to monkeypox involve a coordinated effort by governments, international organisations, and public health agencies to prevent, control, and eventually eliminate the disease. These responses encompass a variety of strategies, including policy development, public health campaigns, surveillance, and reporting systems. This section will delve into how these entities are addressing the challenges posed by monkeypox and the importance of their work in safeguarding public health.

HOW GOVERNMENTS AND ORGANISATIONS Are RESPONDING

Governments and organisations around the world have taken decisive action to combat the spread of monkeypox. These responses are tailored to the specific needs and challenges of different regions, particularly where the virus is endemic or where outbreaks have occurred.

1. POLICY DEVELOPMENT AND IMPLEMENTATION

• **National Health Policies:** Governments in affected countries have developed and implemented national health policies focused on preventing and controlling monkeypox. These policies often include guidelines for the diagnosis, treatment, and prevention of monkeypox, as well as protocols for managing outbreaks. Governments also work to ensure that healthcare systems are adequately equipped to handle cases of monkeypox, particularly in terms of available resources and trained personnel.

• **International Collaboration:** Many governments collaborate with international organisations, such as the World Health Organization (WHO) and the Centers for Disease Control and Prevention (CDC), to align their policies with global standards and recommendations. This collaboration is essential for sharing information, resources, and best practices, particularly in responding to cross-border outbreaks.

2. INTERNATIONAL ORGANISATIONS AND SUPPORT

- **World Health Organization (WHO):** The WHO plays a leading role in coordinating global efforts to combat monkeypox. It provides technical support, guidelines, and resources to countries affected by the virus.

The WHO has also conducts research and monitors the global situation, issuing alerts and recommendations as needed.
- **Centres for Disease Control and Prevention (CDC):** The CDC works closely with governments and public health agencies around the world to provide expertise in disease surveillance, outbreak investigation, and response. The CDC also offers training and resources to healthcare workers and public health officials in affected regions.
- **Non-Governmental Organisations (NGOs):** NGOs play a crucial role in supporting government efforts, particularly in low-resource settings. They often provide medical supplies, support vaccination campaigns, and assist in public health education and awareness efforts.

3. EMERGENCY RESPONSE AND OUTBREAK MANAGEMENT

- **Rapid Response Teams:** In the event of an outbreak, governments often deploy rapid response teams to the affected area. These teams, composed of healthcare workers, epidemiologists, and public health officials, are responsible for containing the outbreak, treating patients, and preventing further spread.
- **International Aid and Support:** During large outbreaks, international aid may be mobilised to support affected countries. This can include financial assistance, medical supplies, and personnel. For example, during the 2022 monkeypox outbreak, various countries and organisations provided support to affected regions, helping to manage the crisis.

PUBLIC HEALTH CAMPAIGNS AND AWARENESS EFFORTS

Public health campaigns and awareness efforts are critical components of the global and local responses to monkeypox. These initiatives aim to inform the public about the risks of monkeypox, promote preventive measures, and encourage early detection and treatment.

1. PUBLIC HEALTH CAMPAIGNS

- **Community Education:** Public health agencies and NGOs often run community education campaigns to raise awareness about monkeypox. These campaigns focus on educating people about how the virus spreads, the symptoms to look out for, and what to do if they suspect they have been exposed. Education materials, such as posters, pamphlets, and videos, are often distributed in local languages and tailored to the cultural context of the target audience.
- **Media Outreach:** Media campaigns, including radio, television, and social media, play a vital role in disseminating information quickly and widely. During outbreaks, public health authorities use these platforms to provide regular updates, share prevention tips, and combat misinformation.

2. TARGETED AWARENESS EFFORTS

- **Healthcare Workers:** Healthcare workers are often the first line of defence against monkeypox, making it crucial that they are well-informed and equipped to handle cases. Training sessions and workshops are commonly held to ensure that healthcare professionals are up-to-date on the latest guidelines for diagnosis, treatment, and infection control.

•**At-Risk Populations:** Awareness efforts are also targeted at populations at higher risk of contracting monkeypox, such as those living in or near endemic areas, people who handle animals, and communities with limited access to healthcare. These efforts include providing vaccinations, distributing personal protective equipment, and conducting community health talks.

3. SCHOOL AND WORKPLACE INITIATIVES

•**Educational Programs:** Schools and workplaces can be important venues for raising awareness about monkeypox. Educational programs aimed at students, teachers, and employees can help spread accurate information and encourage healthy practices, such as regular handwashing and avoiding contact with sick individuals.
•**Preparedness Plans:** Workplaces and schools may also develop preparedness plans in the event of an outbreak, ensuring that they can respond quickly to prevent the spread of the virus within their communities.

THE IMPORTANCE OF SURVEILLANCE AND REPORTING

Surveillance and reporting are fundamental to the effective management of monkeypox. These systems allow public health authorities to monitor the spread of the virus, detect outbreaks early, and respond swiftly to prevent further transmission.

1. SURVEILLANCE SYSTEMS

•**National Surveillance**: Countries with endemic monkeypox often have surveillance systems in place to monitor cases and track the spread of the virus. These systems rely on healthcare facilities to report suspected cases, which are then investigated and confirmed through laboratory testing.

•**Global Surveillance Networks:** International organisations like the WHO and the CDC maintain global surveillance networks that collect and analyse data on monkeypox cases worldwide. These networks help identify trends, detect outbreaks early, and coordinate international responses. They also provide valuable data for research and the development of new prevention and treatment strategies.

2. REPORTING MECHANISMS

- **Mandatory Reporting:** In many countries, healthcare providers are required to report cases of monkeypox to national health authorities. This mandatory reporting is crucial for tracking the virus's spread and ensuring that resources are directed to where they are needed most.
- **Public Reporting and** Transparency: Transparency in reporting is essential for maintaining public trust and ensuring that communities are informed about the risks of monkeypox. Governments and public health agencies often publish regular updates on case numbers, outbreak locations, and response efforts.

3. DATA SHARING AND RESEARCH

- **Collaborative Research:** Data collected through surveillance and reporting systems is invaluable for scientific research on monkeypox. By sharing data across borders, researchers can study the virus's behaviour, identify risk factors, and develop new vaccines and treatments. Collaborative research efforts have already led to significant advancements in understanding monkeypox and improving response strategies.
- **Technology and Innovation:** Advances in technology, such as the use of geographic information systems (GIS) and mobile health apps,

have enhanced the ability to track and respond to monkeypox outbreaks. These tools enable real-time data collection and analysis, helping public health authorities to make informed decisions quickly.

Global and local responses to monkeypox involve a combination of government action, international collaboration, public health campaigns, and robust surveillance systems. These efforts are crucial for preventing the spread of the virus, protecting public health, and ultimately eradicating monkeypox. The ongoing commitment of governments, organisations, and communities to these efforts will determine the success of our fight against this emerging infectious disease. By staying vigilant, informed, and prepared, we can effectively manage monkeypox and reduce its impact on global health.

THE FUTURE OF MONKEYPOX

As monkeypox continues to emerge as a global health concern, it is essential to anticipate the future landscape of this disease. Addressing potential risks and challenges, advancing research and development, and preparing for future outbreaks will be key to managing and ultimately overcoming monkeypox. This section explores these critical areas and outlines the steps needed to safeguard public health.

POTENTIAL RISKS AND CHALLENGES

The future of monkeypox is fraught with several risks and challenges that must be addressed to prevent widespread outbreaks and minimise the impact on global health.

1. INCREASING HUMAN-ANIMAL INTERACTIONS

- **Deforestation and Habitat Encroachment:** As human populations expand and encroach on wildlife habitats, the likelihood of zoonotic diseases like monkeypox crossing over to humans increases. Deforestation and urbanisation are leading to more frequent interactions between humans and animals, raising the risk of transmission.
- **Wildlife Trade:** The illegal wildlife trade continues to be a significant challenge, as it facilitates the movement of animals that may carry the monkeypox virus. Without strict regulation and enforcement, the trade of wild animals for food, pets, or traditional medicine could lead to new outbreaks in areas where the virus was previously unknown.

2. GLOBAL TRAVEL AND URBANISATION

- **Spread through Global Travel:** In our increasingly connected world, global travel can accelerate the spread of monkeypox, turning local outbreaks into international public health emergencies. Travellers from endemic or outbreak areas may unknowingly carry the virus to other regions, where it could spark new clusters of infections.
- **Urbanisation:** The growing trend of urbanisation, particularly in developing countries,

poses a challenge for controlling monkeypox. High population density in urban areas can facilitate the rapid spread of the virus, making it more difficult to contain outbreaks.

3. EMERGING VARIANTS AND RESISTANCE

•**Virus Mutation:** Like other viruses, monkeypox has the potential to mutate over time, leading to new variants that may be more transmissible or resistant to existing treatments. Monitoring these mutations and adapting response strategies accordingly will be critical for maintaining control over the disease.

•**Vaccine and Treatment Resistance:** As monkeypox evolves, there is a risk that the virus could develop resistance to current vaccines and treatments. This would complicate efforts to manage outbreaks and protect vulnerable populations, highlighting the need for ongoing research and development.

4. PUBLIC HEALTH INFRASTRUCTURE

•**Limited Resources in Low-Income Regions:** Many of the countries most affected by monkeypox have limited public health infrastructure and resources, making it challenging to respond effectively to outbreaks. Strengthening these systems is essential for improving disease surveillance, diagnosis, and treatment.

•**Vaccine Distribution:** Ensuring equitable access to vaccines is a significant challenge, particularly in low-income regions. Disparities in vaccine distribution could leave vulnerable populations unprotected and contribute to the continued spread of monkeypox.

RESEARCH AND DEVELOPMENT IN MONKEYPOX PREVENTION AND TREATMENT

Advances in research and development are crucial for improving prevention, treatment, and overall management of monkeypox. This includes the development of new vaccines, treatments, and diagnostic tools.

1. VACCINE RESEARCH AND DEVELOPMENT

•**Improving Existing Vaccines:** While current smallpox vaccines, such as ACAM2000 and Jynneos, provide some protection against monkeypox, there is ongoing research to improve their efficacy and safety. This includes developing vaccines that offer longer-lasting immunity and fewer side effects.

- **Developing New Vaccines:** Researchers are exploring the possibility of creating new vaccines specifically designed to prevent monkeypox. These vaccines would target the monkeypox virus more directly and could be more effective in preventing infection and controlling outbreaks.
- **Vaccine Accessibility:** Research is also focused on developing vaccines that are easier to store, transport, and administer in resource-limited settings. This could help improve vaccine coverage in areas most at risk of monkeypox outbreaks.

2. TREATMENT INNOVATIONS

- **Antiviral Drugs:** The development of antiviral drugs that specifically target the monkeypox virus is a key area of research. While some antiviral treatments, such as tecovirimat (TPOXX), are currently used to treat monkeypox, there is a need for more effective and widely available options.
- **Symptomatic and Supportive Care:** Research into improving symptomatic and supportive care for monkeypox patients is ongoing. This includes developing new methods for managing pain, preventing complications, and supporting recovery, particularly for severe cases.
- **Monoclonal Antibodies:** The use of monoclonal antibodies as a treatment for monkeypox is another

promising area of research. These lab-made antibodies could help neutralize the virus and reduce the severity of the disease.

3. DIAGNOSTIC ADVANCEMENTS

•**Rapid Diagnostic Tests:** Developing rapid and accurate diagnostic tests for monkeypox is critical for early detection and containment. These tests would allow healthcare providers to quickly identify and isolate infected individuals, reducing the risk of further transmission.

•**Point-of-Care Testing:** Point-of-care testing, which can be performed outside of traditional laboratory settings, is another important area of research. These tests would be particularly useful in remote or low-resource areas where access to diagnostic facilities is limited.

4. SURVEILLANCE AND DATA ANALYSIS

•**Genomic Surveillance:** Genomic surveillance of the monkeypox virus helps researchers track mutations and identify new variants. This information is vital for adapting vaccines and treatments to remain effective against the evolving virus.

- **Data Sharing and Collaboration:** Continued collaboration and data sharing among researchers, public health agencies, and governments are essential for advancing our understanding of monkeypox and improving response strategies. Open access to data can accelerate research and lead to more effective solutions.

PREPARING FOR FUTURE OUTBREAKS

Preparation is key to mitigating the impact of future monkeypox outbreaks. By enhancing global and local preparedness, we can reduce the risk of large-scale outbreaks and ensure a swift, coordinated response.

1. STRENGTHENING PUBLIC HEALTH SYSTEMS

- **Capacity Building:** Investing in public health infrastructure, particularly in countries where monkeypox is endemic, is essential for improving outbreak response. This includes training healthcare workers, improving laboratory capabilities, and enhancing disease surveillance systems.
- **Emergency Preparedness Plans:** Governments and public health agencies should

develop and regularly update emergency preparedness plans for monkeypox. These plans should outline procedures for detecting and responding to outbreaks, including the deployment of rapid response teams and the distribution of medical supplies.

2. ENHANCING GLOBAL COLLABORATION

•**International Cooperation:** Global collaboration is vital for addressing monkeypox, as the disease knows no borders. Governments, international organisations, and NGOs must work together to share information, resources, and expertise in combating the virus.

•**Cross-Border Coordination:** Effective coordination between neighbouring countries is essential for managing outbreaks that cross borders. Joint surveillance efforts, shared emergency response plans, and coordinated public health campaigns can help contain outbreaks and prevent international spread.

3. PUBLIC AWARENESS AND EDUCATION

•**Ongoing Public Education:** Continuous public education efforts are necessary to keep communities informed about the risks of monkeypox and the steps they can take to protect themselves. Public health campaigns should focus on promoting preventive measures, recognizing symptoms, and encouraging prompt medical attention.

•**Combating Misinformation:** In the digital age, misinformation can spread rapidly, leading to fear and confusion. Public health authorities must proactively address misinformation by providing clear, accurate, and timely information about monkeypox through various media channels.

4. VACCINE AND TREATMENT STOCKPILING

•**Strategic Stockpiles:** Governments and international organisations should establish strategic stockpiles of vaccines, antiviral drugs, and medical supplies that can be quickly deployed during an outbreak. This ensures that resources are available when and where they are needed most.

•**Equitable Access:** Ensuring equitable access to vaccines and treatments is crucial for protecting vulnerable populations and preventing the global spread of monkeypox. International efforts should focus on making these resources available to all countries, regardless of economic status.

The future of monkeypox presents both significant challenges and opportunities. By addressing these potential risks, advancing research and development, and enhancing preparedness, we can better manage and ultimately control monkeypox outbreaks. A coordinated global response, coupled with local efforts, will be essential in safeguarding public health and preventing the spread of this emerging infectious disease. With continued vigilance, innovation, and collaboration, the global community can overcome the threat of monkeypox and ensure a healthier future for all.

CONCLUSION

Monkeypox presents a multifaceted challenge that requires a concerted effort from global and local entities to effectively manage and control. By understanding the key takeaways and following recommendations, individuals and communities can better prepare for and respond to the threat of monkeypox. Staying informed and adopting protective measures are crucial steps in safeguarding public health and preventing the spread of this disease.

KEY TAKEAWAYS AND RECOMMENDATIONS

1. UNDERSTANDING MONKEYPOX

• **Nature of the Virus:** Monkeypox is a zoonotic virus that can spread from animals to humans and between humans. Recognizing its symptoms, transmission methods, and potential complications is essential for effective management.

• **Emerging Threat:** Although monkeypox is not as widespread as some other viral diseases, its potential to cause outbreaks and its zoonotic nature make it a significant public health concern, especially in regions with high-risk factors.

2. PREVENTION IS THE KEY

• **Animal-to-Human Transmission:** Avoiding contact with wildlife and practising good hygiene can reduce the risk of monkeypox transmission from animals. Measures include safe handling of animal products and controlling rodent populations.

• **Personal and Community Measures:** Implementing personal protective measures, such as regular handwashing and avoiding close contact with infected individuals, helps prevent human-to-human transmission. Community-wide efforts, including public education and vaccination campaigns, are also crucial.

3. TREATMENT AND MANAGEMENT

• **Current Options:** While there are effective treatments and supportive care available for monkeypox, ongoing research into new treatments and vaccines is necessary to improve outcomes and response strategies.

•**Managing Complications:** Awareness of potential complications and prompt medical care can improve recovery and reduce the severity of the disease.

4. RESEARCH AND DEVELOPMENT

•**Advancements Needed:** Continued research into vaccines, antiviral drugs, and diagnostic tools is critical for better prevention and treatment of monkeypox. Investment in these areas will enhance our ability to manage and control future outbreaks.

5. GLOBAL AND LOCAL COLLABORATION

•**Coordinated Efforts:** International collaboration and local preparedness are key to effective outbreak management. Strengthening public health systems, enhancing surveillance, and ensuring equitable access to resources are essential components of a comprehensive response.

1. STAY UPDATED ON PUBLIC HEALTH INFORMATION

•**Follow Reliable Sources:** Keep up with updates from reputable sources such as the World Health Organization (WHO), Centers for Disease Control and Prevention (CDC), and national health authorities. These organisations provide accurate and timely information about monkeypox, including outbreak updates and preventive measures.

•**Monitor Local Health Advisories:** Pay attention to local health advisories and alerts, especially if you live in or travel to areas affected by monkeypox. Local health departments often provide specific guidance relevant to your region.

2. PRACTISE PERSONAL HYGIENE

•**Handwashing:** Regularly wash your hands with soap and water, especially after touching animals, handling food, or coming into contact with potentially contaminated surfaces.

•**Avoid Contact:** If you suspect you have been exposed to monkeypox or are experiencing symptoms, avoid close contact with others and seek medical advice promptly.

3. EDUCATE YOURSELF AND AND OTHERS TO KNOWN

•**Learn About Symptoms:** Familiarize yourself with the symptoms of monkeypox, such as fever, rash, and swollen lymph nodes. Early recognition can lead to quicker diagnosis and treatment.

•**Share Information:** Educate friends, family, and community members about monkeypox and how to prevent its spread. Awareness and understanding are key to effective prevention.

4. GET VACCINATED

•**Vaccination Availability:** If you are in a high-risk group or live in an area where monkeypox is prevalent, consider getting vaccinated with available vaccines. Vaccination is a crucial tool in preventing infection and controlling outbreaks.

•**Consult Healthcare Providers:** Discuss vaccination options with your healthcare provider to determine what is appropriate for your situation and risk level.

5. BE PREPARED FOR OUTBREAKS

•**Emergency Plans:** Have an emergency plan in place for yourself and your family, including how to access medical care if needed. Be aware of local resources and health services available during an outbreak.

•**Follow Public Health Guidelines:** Adhere to any public health guidelines or measures implemented during an outbreak, such as quarantine or isolation protocols.

In conclusion, addressing the challenges posed by monkeypox requires a comprehensive approach that combines prevention, treatment, research, and global collaboration. By staying informed, practising preventive measures, and supporting public health efforts, individuals and communities can play a vital role in managing and controlling monkeypox. Ongoing vigilance, education, and preparedness are key to safeguarding public health and minimising the impact of this emerging infectious disease.

THANK YOU

I extend my deepest gratitude to everyone who contributed to the creation of this book.

To the medical professionals, researchers, and public health experts whose knowledge and dedication inspired every page, thank you for your unwavering commitment to safeguarding our health.

To my family and friends, your constant support and encouragement have been my guiding light throughout this journey.

And to every reader, thank you for taking the time to educate yourself on this important topic. Your commitment to learning and protecting your community is truly recoverable.

Leaving a review is simple and only takes a few minutes. Your support and feedback are greatly appreciated.

Thank you once again!

www.ingramcontent.com/pod-product-compliance
Lightning Source LLC
Chambersburg PA
CBHW070407230526
45471CB00006B/2695